FAST FORWARD

FAST FORWARD

PETER PORTER

Oxford New York
OXFORD UNIVERSITY PRESS
1984

Oxford University Press, Walton Street, Oxford OX2 6DP
London New York Toronto
Delhi Bombay Calcutta Madras Karachi
Kuala Lumpur Singapore Hong Kong Tokyo
Nairobi Dar es Salaam Cape Town
Melbourne Auckland
and associated companies in
Beirut Berlin Ibadan Mexico City Nicosia

Oxford is a trade mark of Oxford University Press

British Library Cataloguing in Publication Data
Porter, Peter
Fast forward.
I. Title
821 *PR*6066.073
ISBN 0–19–211967–2

Library of Congress Cataloging in Publication Data
Porter, Peter
Fast forward
I. Title.
*PR*9619.3.*P*57*F*3 1984 821 84–1093
ISBN 0–19–211967–2

Printed in Great Britain by
J. W. Arrowsmith Ltd.,
Bristol

For Clive James

ACKNOWLEDGEMENTS

Acknowledgements are due to the editors of the following periodicals in which some of these poems first appeared: *Ambit*, *Apple*, *Aquarius*, *Encounter*, *English in Media Studies* (Grampian) *Island*, *London Magazine*, *London Review of Books*, *New Poetry 7* (Arts Council Anthology), *New Statesman*, *Poetry Book Society Christmas Supplement 1982*, *Poetry Review*, *Scripsi*, *Strawberry Fare*, *Times Literary Supplement*. Two poems were included in an Anvil Press pamphlet, *The Animal Programme*, and three others in the Leeds University pamphlet, *The Run of Your Father's Library*. 'Going to Parties' was printed in *Philip Larkin at Sixty* (Faber 1982), edited by Anthony Thwaite. Some other poems were first read in radio programmes broadcast by the BBC.

Again, financial support from the Literature Board of the Australia Council was of great help to me while these poems were being written. I am grateful to the Council for its generous assistance.

CONTENTS

At the Porta Humana

We, the intelligent,
who print ourselves with words,
dream of a race as natural as snails
who talk by walking on footpaths
and whose clatter shines through
clambering of customs.

There, behind the words,
those artful façades for which
so many sacrifices are demanded,
even to the agony that hardens,
sits Stuffy, old signaller
of sickness, always wanting things.

To be loved, to be lovable,
and to print 'not negotiable'
on moments of high illumination—
that's his uncreating touch,
turning perfection back
into personal dots and quavers.

Naturally, real truth
in its comely self-protection
shuns this guardian of drabs:
it takes holidays
among tragic brochures,
even to the pliant madhouse.

The galleries, the gardens
fill with its humanist harvest,
its belvederes are impaled on beauty.
Stuffy sits sharpening pencils,
writing explanatory letters
about love to the psychiatrists.

And the great gong sounds,
ordering, '*forget your fear of faces,*
of the inexplicable, men in the lift
with too much loathing—
populate the prose-world,
inseminate the versicles.'

But can there be a time
for plainness in this jazzle?
Can the plates and arms of fear and love
keep the species talking? Which words
will come through air unbent,
saying, so to say, only what they mean?

The Decline of the North

Round the house, among the ruined cars
And pick-ups, where the armoured lizards
Shelter in a tyre, five dogs are chained,
Five kelpies on a statutory watch.
Decibels of silence fill the day
When belling dogs and creek are tired at heart:
A kingdom comes with dust, a slaughtered sheep
Hangs from a river oak, the text of life.

Your heresy is in new starts, that wheels
And economics travel latitudes
Across the bays of hope. Home might be anywhere,
As otherwise perhaps as in a Jute Town's
Darkened warehouses abutting water
And the smell of sugar burning—that or glass
Receiving seafood pizza, hands at love
And novels with the rigor of the hour.

The Flock and the Star

They stepped through the gate of life,
They moved by emergency
To the exact place of delight
There by the instinct tree.

Palmer stood them in gold
And Blake invented truth
To supplement their world,
A prison without a roof.

They are the people we know
We have been or will become:
To see them best you draw
Night and a star from the sun.

They huddle because they have no
Purpose and yet are alive.
Perhaps the beauty they see
Is why they are a tribe.

Take to this picture, God,
Consider the thing you are—
The unguideable flock,
The painstaking star.

The Missionary Position

Since those first days it has been like this with Tellus,
That the stiff sky is always on the point of breaking
And intervention, above the roof of a shearer's hut
Or in a shower of gold, is what we may expect
Of such pent-up presences. Then there comes that
Augustinian fairness which gives us metaphors
Appropriate to our condition. Upon the roads
After a drenching resurrection, or eating in darkness,
The abandoned followers are shown a sign so
Comforting it sets as a star upon their flag,
A dog recognizes its master by a tone of tail,
The pools fill up with abstractions of the Spring.
The Little Soul itself, natured and then flung upon
The world, revisits cabins of its gloom, kiosks
Where marvellous things matured. Among the queueing gods
Long Service Veterans sigh for ordinariness,
To bring the brillant seed of the future down
To an average hearth, there among the rugs and rules
Of generations, removed from graphs of passion,
Serve their chosen companionship face to face,
The decency and boredom of diurnal love.

Analogue of Helen

All ladies should have one,
an eidolon for days
they kicked over the traces
so they can say
'I was here all the time in spirit
which makes the presence
of my physical body in that other place
of little consequence.'

Euripides puts it the other way round,
all Paris got was a light show,
but we are more practical
and believe in equality.
Let it be herself she disposes
not just for adultery
or interesting changeableness
but to reinforce her
exquisite readiness
to meet each moment as absolutely
alive: then she will bring
to long-postponed quarrels
some of that wide-ranging
understanding her husband displays.

And let there be signs
that the usefulness of lies
is fully understood
by civilized persons.
What you write down in love
is never untrue; the paper
grows old around the words
and any truth-loving ego
would fight a ten-years' war
in the family house to keep it so.

Eyes look at the world as emperors,
even their soft return at evening

is feudal. Where they cannot enter,
touch too is kept out.
Perhaps oddity alone will tame
baffled possessiveness—the boy
Auden and his mother
singing and playing the *Liebestod* together.
When the neurons in the brain
are tucked-up for the night
a man may have Helen to himself.

Jumping to Conclusions

The most unexpected things,
the way girls in the street suddenly
come to resemble Pontormo's women
and behind a frond at the pizza palace
Arethusa's fountain frothing,
the map of the sexual republic spread.
Let them race to surf or tennis,
the true corona is in books:
later, at a party, tell a blonde
how the castello was built
to terrorize the subjects of the Duke,
and when first light fizzes in your bed
console yourself that nothing but bad
literature ever happens in real life.
Nor is this a corseting or structuring,
writing essays and not poems.
Rather it's divine displacement,
the world not having any fairness
our urge to perfectibility
seeks quality in randomness.
Going backwards can be good,
Pope never again attained the maturity
he had at twenty-one: we have to know
more than we can feel, we have
to make heartless aphorisms
out of misery, even our own despair.
Such charming vistas, the river banks
approaching death, the postures of the gods,
Illyrian survival! Doomed to claim
all that we envisage, we startle
watchers on peninsulas
with feats of simply seeming,
the form is in the summoning.
Those monkeys typing encyclopaedias
seem devotional, even tactful,
they could be working on
a New Book of the Dead. Then to borrow

sweetness for a second from the species,
irritable dreams, jumping to conclusions
among the delegates with cardboard badges,
greet the future at a soapstone villa,
the history of the world upon your tongue.

Poem Employing Words from an Article on Combat Strategy

In the garden of our first Summer's day,
the suspension of disbelief is shown
as something to do with taking the strain,
metaphor solidifying itself
in persons once again, and here comes
the purest attrition mind-set
concluding with a slim girl in tailored jeans
asking for *Disque Bleu* in a Dumfries accent.
About her buffets the wind's body-count,
more than whole numbers of the sun,
evolution proved on deck. But she must die,
worse, must age. The General Staff cannot agree
what would fit her for a high tech. fate,
what little and expensive love
might get her flying. Sex and bewilderment
allay the afternoon, gold-plating
for the mind's most complex model.
A pity we cannot use the phrase 'come out'
of her, since that is what she's done
in emblems of the Summer, no more
a hangar-queen, with all her polished parts
stripped down, but manifest in service,
a butterfly upon the wing.
The deterministic crowd
parts for her as she slips sideways
into Cowgate, back to its beer,
her truck-kills garnered in the gloom
of ever-open *post meridiem*.
Another little gismo
on reality, says the poet, accessory
before the fact of helpessness.

Cleaning the Picture at the Edges

At the Retrospective, they were full of
transferred clarity, voices explaining
that elegance is how the lecturer keeps
ahead of his bright students—his rust chimneys,
cyanide sky, Sickert taxi skidding
seen blooming alongside portraits
judged novelistic, a bend in the colon
of time/space/taste. A reviewer added
this damaging gloss: Rembrandt's grandeur
is made the subject of the caution
of a pale contemporary, thus it dies
in neatness. But he didn't ask
whether R's greatness wasn't a seeking-out
of peace, its exhaustion of itself
a quest for perfection, tidy in its way
as the neatest poem. Tomorrow Rembrandt
would have to start again, there is no end
to *terribilità* (let us for once take
Michelangelo off duty). So it is with us,
we are sure that Bach has more in common
with Shakespeare than with Gödel
but we should be careful of the books we read.
Was there not a dream where someone
rather like a poet introduced himself
as 'Anon Sequitur'? Dreams admit no jokes,
it was a warning. Then the quavers
pushed the Jordan through its banks again
and calmness separated from the dark.
Where I am sitting I have put the whole past
into a room although I know the papers
at my feet are no more than the rubbish
which floats at night to corners of the Baths.
A page from a memorandum pad
supplied by a pharmaceutical firm,
Sadlers Wells in 1957, a hopeful letter
bringing Leeds to London—I must start
to file it all away. Pain will set me on,

it wants its peace as well. Those parts of my brain
which have escaped the claws of alcohol
are setting up a federation: what
have you done to make our future certain?
Cleaned the picture at the edges, whispered
to the queue of frightened memories,
'It doesn't matter where the semen goes'—
Now I shall look into detail
for a census of good style, assigning
images to their stations on the Overground.

Clipboard

Just this from a life and house of plenty,
The racks of records, the books unread and read,
(Buy them and they cannot persecute you)
Cats dead, one living, wife a sheet of contacts,
Children flighting of their own volition,
Television filtering among wine glasses,
The horrible telephone biting its tongue,
A few out-of-place tropical plants—
Only this to offer the dungareed gateman,
A clipboard of papers, roster of reality,
Off-prints of dreams, the documents of an end.

The bottom ones so curly yellow and the type
Old-fashioned. They go with gardens in the heat,
A university of touch under the house
And fields of strangers circling on a lawn.
The middle ones official, stamped with what
Might be insignia of failure or success,
So interchangeable their world. The ones on top have splashes
You might use to authenticate
A legacy to libraries—coffee, wine or semen.
Whatever their value or significance
These have been gathered to present you here.

They are your hieroglyphs, your flying signs
That wing the soul under the roofs of heaven.
How much more glowing in the dying light
Might seem those towering and objective works
You never saw or mastered, that way of truth
Opaque to all biography. The sonatas, genre scenes,
Odes to Demeter are not included. Paper betrayal,
Words unredeemed by anything but death,
Such to be handed to the unattentive porter
At the gates. You lived on these poor promises,
Now they must be your friends and witnesses.

The Arbitrary Abrolhos

After indifference,
a survey of vulnerability—
a liftdriver with Liszt's keyboard
in his smile, the Festival of Open Desks,
the arbitrary abrolhos!
What evidence
will I file against her?
Love is a fluke you might make
evolutionary by faith. Till then
dreams and logic. The grass beneath your feet
is a station of fond tombs,
the bees are fed on memory.
'I have not worn
my wedding ring since meeting you,
absolutely not a coincidence.'
Which of her many faces will
the sea-shouldered goddess wear
when she summons me for judgement?
Shadows breed shadows,
corners for dead fears, her picture,
reclining on the royal balustrade
or sharing a gin-and-tonic
with a sculpted harpy. Now to retire
like Diocletian from the mess
of a mad empire, an end to naming—
She is the image
silently controlling valency.

Doll's House

Against the haunting of our cats,
Shy raids by children visiting, it stays
As truthful as the willow flats
Which blocked her days.

Its owner slammed the door and fled
Like Nora to the liberal hinterland.
What could resite that jostled bed?
No grown-up hand.

The miniature hoover lies
Brim-full of dust, the chest-of-drawers gapes;
On holidays a sobbing tries
To fluff the drapes.

And now to play at house you need
Another sort of house inside your head
Where duty states you soothe and feed
The plastic dead.

Her children have outgrown it too,
But do they hear the twisting of the key,
Entail their ruined space in lieu
Of charity?

Love, orderer of dolls and towns,
Has Liliputianized the scale of pain,
So the wide adult eye looks down,
Bereaved again

Of esperance, the childhood flush,
And has no passage into afternoons
But through diminished doors and hush
Of darkened rooms.

Comedy Lies

Thinking of the different sorts of loneliness—

Adolescence and not to be a student
amid the glaring competence of youth—

Marriage and wheels turning in the clock-
bound passages of kith and kitchen—

Holidays with lines at midnight
playing back from hours on crumbling beaches—

Knowledge that however serious the world
its terrible moments return as jokes—

Wonderful for those poets who keep away
from meaning, living somewhere better—

Forgive me, my love, but I can remember
the poor things I wrote when I was trying—

That our private language should look cold
when all that's left of it is public

Is not surprising. Every answered dream
is merely editorial in sunlight—

Thus Hardy's point: 'Tragedy is true guise',
feelings never appearing out of uniform—

But how could we, such amateurs,
afford a panoply? We were laughing unto death—

So I pile beside my hand those books
which hold my lies and I hope you'll say your lies—

All I hear are heavy lines of diphthongs,
they tell of two people (you and I) in touch—

All sorts of loneliness become the same—

Venetian Incident

I take for my sermon a Sunday
in Venice, walking from the dazed hotel
along the shortest way
to the Salute, and then, as so often, propelled
by my Britannic bowels, forced to take
a red-white-and-blue trip back
to our room, and you, rather than make
a fuss, said you'd meet me at the Anglican Church
on Campo San Vio (I found
the name in the Pisan pages of Pound
and it moved me more than I guessed
such a glossing reference could)—
My intestines are a species' research,
but soon enough I entered, a guest
of the Low Church sort, in the House of God
and saw you cheerfully raising and lowering,
along with some seven people and a dog,
in a ritual just short of Rome
and the priest, camply flowering
through the familiar words. You were home,
a devoted unbeliever but English and real:
I a frightened sceptic
ever willing to make a coward's deal
but till then too stiff and quick
to be at ease in High Church mysteries.
'I know my way through Holy Communion
by heart, ever since that awful Father John
taught us at Elmhirst.' Your words,
your dispensation. But God was pleased.
How do I know? Because, even if he doesn't exist,
he likes us to try to belong,
and though you were not relaxed in his world
and needed by lunch to be pissed
your lightness at living and dying expressed
Creation's plain tenure, the warrant of birds.
I saw then that those who cling to their life
are death's real retainers, that the Mass

mumbled and bobbed through is counted a pass—
I could outlive my wife
but never be natured into the space
which she proprietored.
It will never be sufficient to speak of
the peace following after;
some text must be found to remove
the permanent discord
which flesh sets on flesh. Whence comes the need
for punishment, as native to liberal souls
as to Savonarolas? Our minds breed
a cancer of starting—from God or Black Holes
we surge into certainty and thence
to eternity, dying and dying and dying
and always the wrong side of the fence.
It was joy then to pause on the path
at things coming round once more,
some words and some gestures, familiar and hollow,
a welcoming door,
the voice of our language and our sort of dogs,
expatriate ladies that cats learn to follow,
a chance of forgiving
with Venice behind us, like us exhausted
by life, and pleased to be living.

Elegy and Fanfare

When that cry which broke the heavens
intervened in the poet's rage,
the shape of Creation showed itself.
Think of God's large work, for which
armchairs were set out by prophets
and by elders; think this relived
by each new mind encroaching on to life—
Thus the Rilkean spatter of angels,
a glass shaking at sea; and thus
the business of invention,
dispersing feeling's clamour,
our duty to rein melancholy.
    Where in my damaged words
will you surface now to me,
bright with separation
as if eight years' passing
had washed away all pain,
language once more the twisted path
to knowledge? Standing by the headstone
in that neatly fenced new cemetery
in Cambridgeshire, I was fanned
by all your absence, wings intending
degrees of nothing, a pleasant void
one tear would rupture—and no tear came
to swing the afternoon up to my face
or as a postulant to ancient trees
alter the pace of selfishness,
my expedition to Newmarket Races.
    The cry Rilke heard
was never stilled among our fallen gods:
I listened for it through the midday hush
accommodating bees upon the little blooms
of this scarcely half-filled graveyard.
I could not fancy what his angels made,
carved out of air; indeed both past and present
murmured on. 'Whither thou goest
I shall go, thy people shall be my people

and thy God my God.' Leaded windows
enclosing pannelled glass the colour
of sucked sweets, the passion of Ruth
for daily loyalty, and myself
a boy already conscious of being born
out of his time. This connected with sunny
suffering, and spurred by cowardice,
I imagined a welcome from the speaking stone,
words of Popean sylphs, the humid nymphs
of memory, '*In deinem Grab*
will ich mit dir begraben sein.'
This is an elegy,
so forgive my German and the words I put
into your mouth. I need this distance
to proclaim the truth, and I'm the one
not yet secure in perfect night.
No tocsin of towns
will resurrect you from the alphabet,
there is only refrigeration of remembrance,
that rhetoric the poet heard booming
and blooming in himself, most precious
of his properties, as, starred with vigilance,
he looked down the unrecurring light
to death. In his hell his heaven blazed.
Those angels fanfared
once above the Adriatic, then again
in salons of his Middle Europe.
For me they were the English ends of hope
from which I now reluctantly retreat
recalling insolence of a young man's doom,
destined to outlive the only truth
he ever met. In Cambridgeshire
the singing will not stop until the book
of everlasting infancy is closed.
The urn below the soil,
meridian of Meldreth,
is for geographers of finished love.

Where We Came In

I collected my father's possessions,
a half-sovereign case, a gold watch,
fourteen carat only, my mother's rings,
and walked into the breathing sun,
Another heat shield gone.

Fine powder of selfishness along
my upper lip, the time of jacaranda falling,
here where I was born, the cycle
not yet complete but estrangement
Made absolute by time.

I tacked to the car as if I were drunk.
Indeed, I had lost my common sense
of ownership: when inheritance shrinks
to memory and thirty cents in cash,
Who's then the family man?

Yet soon in the bar above the cricket field
I rallied, due to timely punishment.
At last I was alone with incandescence
and did not question the mystery,
The son was now the father.

Dejection: An Ode

The oven door being opened is the start of
the last movement of Rachmaninov's Second Symphony—
the bathroom window pushed up
is the orchestra in the recitative
of the Countess's big aria in *Figaro*, Act Three.
Catch the conspiracy, when mundane action
borrows heart from happenings. We are surrounded
by such leaking categories the only consequence
is melancholy. Hear the tramp of trochees
as the poet, filming his own university,
gets everything right since Plato. What faith in paper
and the marks we make with stencils
when a great assurance settles into cantos.
The Dark Lady was no more than the blackness of his ink
say those whose girl friends are readier than Shakespeare's.
Just turn the mind off for a moment
to let the inner silence flow into itself—
this is the beauty of dejection, as if our unimaginable death
were free of the collapse of heart and liver,
its faultless shape some sort of architecture,
an aphorism fleeing its own words.
Betrayal goes so far back there's no point in
putting it in poems. I see beyond the pyramid
of faces to strong monosyllables—faith, hope and love—
charitable in halcyon's memory, fine days
upon the water and weed round the propellor.
Now all the theses out of dehydration
swarm upon my lids: I was never brave
yet half an empire comes into my room
to settle honey on my mind. Last night
I quarrelled with some friends on politics,
sillier than seeing ghosts, and now this neuro-pad
is dirging for Armenia. Despair's the one
with the chewy centre, you can take your pick.
I listened to misanthropy and had
the record straight. The woman in white,
the lady with the special presents of mind,

may now be on the phone from out of town
just to keep in touch. Think, she usually tells me,
of Coleridge and days in record shops
and all those 'likes' that love is like,
a settlement to put our world in place.
What has the truth done to our children's room?
The toys are scattered, the pillow damp with crying,
chiefly the light is poor and no-one comes
all afternoon: *Meermädchen* of the swamp of mind.
I kept my father waiting, he will know
that the disc, long-playing for however, ends
in sounds of surface, of the hinge and wind,
an average door, a tree against the pane.

Matutinal

This is ramshackle occasional,
the soft centre of a city dweller
watching swinging surgeons
at pedicure upon the plane trees.
It is a sermon of white on blueness
and a blundering at meaning
where we sat outside the restaurant
of dreams. Cover more paper
till my eyes are bright for the dark,
travelling to the mind of flame.
And for those whose flavour
is really afternoon, let light
give them this sinecure of morning,
the quiet flat, new leaves on
the imprisoned plant, a frog-like green,
the restfulness of starting.
I said I was a Puritan
but did not know myself,
disdainful of the cause of living.
I have been rescued momentarily
from all connections,
love perhaps, truth most certainly,
by this substantial missingness
beyond my window: by the 'the'
of itself, the moon-corrected air.

Student Canteen

An arsenal of tilting flesh
Around me with its trays and shine,
Its absolutes of yours and mine,
Impatience's cute micromesh
That keeps the oldest envy fresh—
Is anything more ordinary,
More just the thing you're expecting,
Decent students one weekday?

And yet to license girls with such
Lethal weapons as their own
Faces, voices, tilt of bone,
Characteristic stance, is much
As if the upper goddess touched
Down on earth Endymion-wards,
Not for the sake of beauty's ache
But to ring sex about with swords.

The dream which keeps us comatose
However haggard we become
Is that the moon will think us young,
Go down on us where we repose,
An open excellence like a rose,
And show that hope is truly love—
What we have done concerns no one
Since value is confessed above.

You cannot hold such fancies here
Where youth is massing for the kill
(It doesn't know this is its skill).
The creamy tension of this fear
Lights the coffee, dulls the beer,
Sets constellations in the hair
Of sultry teens of average means,
Their novels unread on a chair.

I come here almost every day
As if I walked from heavy dreams
Into some light supporting scenes,
Metalogue to a Satyr Play.
Discontent won't go away,
It marches miles from class to quad
To parody humourlessly
The chauvinism of a god.

The Biographer Promenades

Consider the terraced living of these cows,
Devoid of soul but denizens of fields
From the Ringback Hills to Divagation Brook—
Pondered at depth, their biographies
Have all the lucid exaggeration of Jesus
From passivity of clover to a Via Dolorosa
Of slatted trucks. Best of all , they typify
My huge good fortune; their lives need me
To make importance—for once the publisher
Will say, 'I am almost in a position
To put nothing but your name upon the jacket.'

This sort of symbolism pleases me.
I am at one with scholars in their freezing texts.
Our exhibits are worthy only because we
Are their custodians. I might stage a search
In one of the better picture libraries
For a shot of a specially scabrous sort
(Like Nietzsche and Rée in the cart at Zürich),
But I would not even hope for more revealing
Information. More means worse. The Devil
Puts his arm round each of us and whispers
'Your secret life is absolutely riveting.'

O but I have seen them! My quintets, sextets, octets,
As I put them in the shafts! On good days,
I drink a narrow glass of water, laced
With lemon, and set out for the Archive
Knowing that the evening brings the raw stuff
Closer, a raffish party at which I act
Pecksniff, and we know, yes we all appreciate,
I'm just kidding—getting in on the ground floor—
But that is where I venture naturally
(The good biographer is part of what he tells),
Upstaging subjects for the public's sake.

Ever since Tacitus it's been like this,
Or perhaps the man who followed Gilgamesh
With pad and pencil. Sorting the plankton, bandar-log
Of followers, girl-friends, pimps, hand-holders,
Mothers, vicars, fellow-undergraduates,
Colluders in the cold. Art can look after
Its own, but men's frail putrid lives
Need advocates. I need a space myself
If not upon Parnassus then on that slope
Appropriate to late democracy
Where words, not bones or faces, are laid down.

Marianne North's Submission

We are born to whiteness
And must mark it with those graphic miles,
Creaturely or vegetable,
Scribbled on the gums of New South Wales,
Finely sententious lines
Though not of God. Such squiggles may seem
Appropriate roads on maps in Heaven. If there is a Heaven
It's because there must be something
We have moved away from. It's hardly heavy,
Only a smoke to emerge from
As if above an orchid-flooded valley,
With a dry sketching-pad, a day ahead of me
And nerves for once indicative.
I cannot remember when I consented
To be born. It was many years after
My entrance on this earth. I was poised
High enough, though what our flesh must do
To be naturally patrician I leave to those
To calculate who find our planet habitable.
So many do, and oh my heart and more my eyes
Go out to unreflective strugglers,
Vines, stamens, steamed-open lilies,
Voluptuous catechizers of the death camp
We call jungle—or the reticulated root,
Shading a scorpion, which knows the hour
To lift leaf systems to the sanded air.
I can bore even myself, working at
My natural wonders. I know as well as you
That this is complex only to the pen
And to my difficult purveyance. I could pack
A camera and put my life's work
Out of business. All my laborious
And loving pictures sometimes seem
Like women's dress materials laid out
In a mercer's pattern-room. I like
All complexities to be of shape
And observation, of pure distinctive essence.

The Higher Creature has simplicities of action
And arrangements called philosophy,
The which they tell me our spirited country
Has brought to their apogee. I put away galoshes
And take out my deepest waders. To any watcher,
Including God, I am a middle-aged lady
Of the dauntless British sort feeling not
Very well in a Burmese swamp—there she goes,
You say, escaping the Victorian Sick Room
But not evading fear. Do you know
My farthest journey has been away from suicide?
Let this lizard fern I'm sketching
Speak for me: 'she serves that filigree
Along the air my svelte and dappled self
Keeps pennanting above our high monsoon,
A single real beyond imagining.'

Cyprus, Aeschylus, Inanition

As we come down to the island, brown
Under blue and leaning on the sea,
The utter gods of festival and burial
Leap into sunshine from necessity
Of history and cement-mixers,
Promising tourists an authentic
Frisson, their unmaskable encounter
With exhaustion, light and vanity.

Whether we are lit by lemon-toothed lamps
Of Aphrodite or switch on bulbs above
A hotel bureau to write out the traveller's
Conceit, 'a polluted Levantine swamp',
We will not disturb those ghosts, history-thin,
Who were old to the Venetians and whose ardour
Has tamed generations of arrivistes.

Moon landscapes of kaolin hills, framed by
Sexual froth where the light-stepping goddess
Walked up the beach like a starlet at Cannes—
So apposite that mosaics should show
Coy Pyramus, reluctant Thisbe, and a lion
Camper than St Mark's, that tee-shirts at Paphos
Are printed with postures of fucking
And the silent amphitheatre sings to the sea.

Waves which bore the Greeks round the world
Have sunk now to evenings in theatres,
More like Noh Plays and The Gang Show
Than Tragedy, capital T. So pretty,
Our barefoot Orestes, such anthropology
As legalistic Athena, corrupt Apollo
Lead the Furies in the State Tattoo!

Where endurance has shrunk the heroes
To midday broadcasts of statistics
And the toing and froing of competitive gods
Must be imagined behind the clean glass

Of archaeological cases, it suits us,
Newcomers withering in history's gale
As the North declines, to put tragedy behind us,
A plain separating no sheep from the goats.

And here we can propitiate the present
Which makes its own deities, to be sorted
By victors from the rubble of contention—
Doric the reinforced-concrete holiday pens,
Ionic this pale commercial Phoenicia,
Corinthian the conference halls—newest gods
Are the cruellest, colliding in the mind.

What comes of the used-up Mediterranean
When rockets point like pines in tundra
Towards the profaned moon? Only an old
Theogony of self whose Pantheon
Is fixed in brain and liver; even birdsong
Has its mensuration, the rules of wordlessness.
Long-buried kings, each wormy iconostasis
Show how we wind down to gods at the end.

The Cats of Campagnatico

Since a harebrained Devil has changed the world
To scenes from a Nature Documentary,
There are those of us who will forever seek
Rational landscapes, dotted with walled cemeteries,
Unquestioned rivers of familiar fords
And an efficient bus from which adulteresses
Alight before the ascent to the neighbour village.
Not that His blocked thumb is absent: those
English families tooting along the scatty road
(Our fifteen-year-old crunching the clutch
of the little Fiat) are outside the cemetery
Before anyone notices the just-widowed blob
At the armorial gates—the regret, the shame
The silence—she at the gardens of death which need
A constant tending, and us hurrying
To lunch at the hydrophilic villa—
The Oldest Presence of All will be well pleased.
Not just a vignette, we reflect, this shadowless day
In Southern Tuscany, more a looking for shades
Which match the petrified intelligence of time:
One sees the small bends which history makes
In the lanes of scarcely-visited villages.
True, this one is in Dante, and that oleander-screened wall
You take for the headquarters of the Carabinieri
Might be an out-station of the Piccolomini,
If only you could remember which is which
Among the towers that mark the lesions of the sky.
Siena is as far away as London; life as far away
As last night's dream whose every promontory
Is in the present. Now, coming through the gate,
The view is a pastoral benediction for those
Who have never lived in Arcadia. *Thank God,*
Grace à Dieu, Gott sei dank— we are
As international as an opera festival,
We who love Italy. We have no home
And come from nowhere, a marvellous patrimony.
Then before the laying of the table in the arbour,

The helpful barefoot girls from good schools,
The gossip and the wine, a sudden vision
Of belonging. The cats of Campagnatico,
Which are never fully grown and have never
Been kittens, will not move for the honking motorist
But expect to be gone round. Thin and cared-for,
Fat and neglected, watchful and hardly seen awake,
Cool-haired in the sun and warm in shadow,
Embodying Nature's own perversity,
They lie on this man-made floor, the dialectic
Of survival. O God, we cry, help us through
Your school of adaptation—between the fur of the cat
And the cement of extinction, there are only
Cypress moments lingering and the long tray of the sky.

A Guide to the Gods

You must recall that here they are intense,
Our gods. Each corner shop may need oblations
So the genius rests content. Kick a dog turd
Leftways to the gutter, but only if it be
Dry and crumbling. Every third display of okra
On the footpath must be stolen from—one finger
Will suffice. Touch your collar points when
Passing electricity showrooms, for there
The ghost of light supports a goddess
Veiled in wrath. Parades of the holy mad
Transferring parcels should be followed
(But judiciously). Bag women and drunks
Have been identified as Furies, but I
Do not wholly credit this. Hurry home before
The postman if you see him near your door:
That letter threatening a visit will not come.
Parse the shorter sentences in Health Food Stores,
Their higher prices are most magical.
Believe me, burst water mains may be
Oracular, but every man of sense resists
Coercion. When they wish to speak to us
The Gods behave with plainness and with modesty.
Next time I'll tell you about Zeus
Haephestos, Artemis and other probabilities,
But that is fiction and not upon our scale.

Santa Cecilia in Trastevere

I found here only the music of exclusion,
gates locked upon flowers and fountain,
a *hortus conclusus* of organized sound
for its mother to rest in when the birds
and children blanched her silence. I had wanted
this best of picture-postcard saints to be at home
when I called, bringing her I thought
greetings from the contrapuntal North,
territories where ears hear differently.

How hard they had to work to kill her—
nearly snuffed out in the calidarium,
hacked in the neck by a swordsman and singing
to the end. But lucky after all to have
this architectural palimpsest for home,
far from those grim and anal catacombs,
the Christian labyrinth. The sun blazed on gates
which kept me out and the garden shone with dust,
a fleck for everything put here to dance.

With Christ and his radiant eschatology
she can have little to do. A cardinal might weep
to behold her bones in golden wrappings, but we
will hear her in the lift of blood as, joyful
for the world she sets before us, patroness,
we trek the dry Janiculum. Her thunder
is the face of beasts, apotheosis of
a natural sweetness, death at the end
sounding a miracle, tempering transcendence.

One of her adepts pronounced that art
posing as religion is the worst vulgarity,
yet what more baffling for the tired pilgrim
than to find himself excluded from a shrine

hallowed by a working saint? Rome and Florence
it seems are always closed to us
but fond Cecilia leaves her door ajar. 'This not
unpleasing Eighteenth Century church' (I quote)
enfolds a darkness round our lightest step.

Unfinished Requiems

How could they come to an end,
With the grateful creator completing
His blots and boredom and writing his
Laus Deo, as he has done so often
And as he knows his Father in Heaven
Would had He someone to look up to?

Symphonies are different, abandoned
Perhaps, essentially themselves
Whatever the state of the manuscript
Or ambivalence of the critic. We'll prove this
By asking you to point to one
Where the long-lost scherzo isn't dull.

That is why there are only two ways out:
Either you plan your requiem for
A hero of the state, and write it all
If your collaborators are sluggish,
Or you welcome The Stranger in Grey
Never doubting who or what he is.

You can escape the danger perhaps
By writing two or more requiems
(If C Minor first, then D Minor later).
Which is the one for yourself? Silly to ask,
They both are, and the liturgy
Is a long letter from you to nowhere.

So even if your fever shouts at you
Halfway through the *lacrymosa*
And your pet canary seems a soul
On Charon's bark, your work
Is no less finished than that of
The shy man bowing towards the Royal Box.

Life is left in middle manuscript
And widows, critics and descendants
Can get it ready for performance
And those warming royalties. Predecessors
Enjoying eternal rest will tell you
The unfinished theme is always bitterness.

Little Harmonic Labyrinth

Come stars and beg of the one star
a progress through the laughing fields
beyond our pink-walled town.
The little monkey on its cushion
brings the priceless gift
of sexual desire. Without this rubbing
luxury there'd be no chasseresse
of envy, nothing but our getting staler
on the avenues of evening.
How dare they share this gift,
these best of lucky solipsists—
'On this soft anvil all mankind was made.'
And the tyrant will,
unrepentent of its mediocrity,
is governor of created things.
The flight from meaning is our magic:
overhead a perfect line of birds
pegged out to dry—the picture shows
where dreams have passed,
trooping to a reborn god.
O captains of your consciences,
the world's a middle sea
washing tearful stories to the shore—
Tell of blushing Psyches
in little breasts and sneakers
bringing serfdom to tomorrow,
reflate the fluffy trees, the cobalt sky,
in allegories of sin
with all our ages snickering in bushes.
Even the guaranteed untalented
have style of their own, our God
has given us immunity
from everything except ourselves.
No wonder I have dreamed
the living and the dead are one,
that out of their congestion
a planet rises which has sounds for air,

whose syntax may be synthesized.
O eyes I cannot meet,
yours, preppy teenage gods,
show me something serious
beyond imagination. When sex dries
all that's left is abstract,
completed outlines without presences.
Find me a star to shine
through the whiteness of the mind.

To Himself

A WORKING FROM LEOPARDI

My exhausted heart
It is time for you to rest.
The final deceit is over, the one
I thought would last for ever.
It is dead, this love is dead
and I am content that with it
dies all hope of fond illusions
and any real desire to harbour them.
Rest now forever, heart,
you have worked too hard,
your every movement comes to nothing
and the earth which moves beneath you
is not worth sighing for.
Bitterness and emptiness compose our world,
there is nothing else; our life is made of mud.
Heart, be quietened now,
you have found your last despair.
To human kind fate has allocated
only dying: scorn Nature then,
the brutal power which rules for misery,
and the vanity of everything that is.

Dis Manibus

IN MEMORY OF CLAUDIUS, DIED AUGUST 1980

Here before me is a space
which should contain you.
I still live in it, ridiculous
this definition of happiness.
You were a cat and humans are designed
to love other special humans.
When they do they try a little.
We never tried with you,
we didn't have to.
We must be decent people
to have been given you for company.

You deserve a very formal poem
in a complex stanza shape
and a cat-flap-banging metre
(but we had no cat flap
and you had to sneak out and play
jokes on us upon the stairs)—
Let's see you take again our teasing love
as you took your narrow world above the street
and pain beside the piano's icy pedals.

God forgive this helplessness of friends.
You died in a box in a car
and love could not get you out.
The theory is that what we write
mimics what we feel
and that's worth a prisoner's laugh.
Sleep, Claudie, where you are.
God's rule-makers will not give
you a soul, but, believe me,
you are no worse off without one.
Winter is upon us, the tree of life
is withering. To settle into death
with your tail at a comfortable curve,
that's style. Short views, few words

and empty evenings now. They are coming
to take me to a marvellous party
and whom do I expect to see? My host—
a big black dazzling ambulatory cat.

A Vein of Racine's

In this vain valley of ambition pleas cannot be bargained.
As in the spent Renaissance, life turns to the ideal,
And thus, distrustful of its place, its plage, the hotting-up
Domain of Demos, it sets its transcendental cooker for
An hour beyond all frenzy to call us to that noble feast,
Interior satori, a dish to put before Self-Emperors.
Sitting in the nylon static of motels, what alexandrines
Have we to repel the public day? Form is a lazy witness
Seeing whatever it finds easiest to see. The classics are brought
  down
To courses on comparative belief. This is not the fault
Of idle teachers; rather such consummate victories were bought
At high rates once by calm custodians in rooms so
Humanistically exact for sunlight and approved debate
That when the trays were cleared and once the mercenaries had
  swarmed
Past tapestry that muffled Tasso, only blows by brutes
Had style. Everything and everyone was clangorous for honesty,
The world became one Salon des Refusées. But do not look
Just at our telemetric world and its refined abstainers:
Consider in its place academies at one of several palaces
Where feudal waters sink into evening gold—what did
Those princes and those poets promise defenestrated Man?
Only to confuse his single fate with that of species,
Blood and the burden of our winsome natures he
Might find a phantom for. Always there was elsewhere,
The Golden Age, the Innocentest Isthmus, a land on
Stalks beyond the eyes of youth. When the Prince had raged
Through statues, tempios and the scrolls of his Isotta
He grafted Nature; he could think of welfare, care for
The nameless many castings his shadow on the square.
They took the hint; love alone will tell all shades of creatures
They are one. Thus came the Proverbs of Democracy—
You will get the broad apotheosis you are waiting for,
You will put the gods inside you and make death their king,
You will know your nervous system a bland oracle—
And yet the gathering storm approaches, it seems that Man,

Tamed by his tapestries to play Actaeon on the plain,
Cannot be redeemed. The youth his girl friend, the warrior
His tall challenge, the poet his obsession in the head,
A programme vulgar as the planet is unveiled:
Power and response, the Myth of Circumstance,
Great men impaled on history—nothing comes between
The goddess and her darkness when she loves, disturbance
Of the common decencies of lust, that moment when she sees
Her prey, about to inherit the innocence of the world.

Going to Parties

FOR PHILIP LARKIN

Truth to experience, to the sombre facts,
  We all believe in;
That men get overtaken by their acts,
That the randy and highminded both inherit
Space enough for morals to conceive in
  And prove the pitch of merit;
Such insight hangs upon the scraping pen
Of the deep-browed author writing after ten.

But there behind him, if he chose to look,
  The ranks of those
Whose sheerest now is always in a book
Are closing; yes, he's broken up some ground;
It lies about in other people's prose,
  Great graces that abound—
Meanwhile, incorrigibly, people seem
To write their own existence from a dream.

Perhaps it makes him think of earlier days
  When parties beckoned,
Quartz studs glittering in a bank of clays,
And he'd set out, though apprehensive,
Hoping this time to come in first, not second,
  Ready to really live,
Only to find that life which offers chances
Ignores the sitters-out and picks the dancers.

Yet he got something there. How to enjoy
  Expansive moments,
How it must stun the gods to be a boy
Who will not bear the cup, how unasked
Guests act prosecution and defence
  (Who was the man in the mask?)
And how exhilirating when alone
To know those dandies walk on stilts of bone.

Tributes then; the party isn't over,
  A few guests linger.
From heartland England on to distant Dover
People are shutting doors they know too well
And following their feet to hope, the bringer
  Of several shapes of Hell,
Of time, experience, and all we use
To make art of a life we didn't choose.

The Hedonists Rehearse the Uprising

In two hours it will be Fasching
and our watches will tell us
we are having a good time.
Misery, say those returning from
the temple where the scanner
reads the lump, is as palpable
as sunshine.
                    The countrymen of God
are setting out the sails, the oars,
the sandwiches, or to some,
who like to place their bets both ways,
the stomach pumps.
                    Slaves to duty,
as much at fifty as fifteen,
the hedonists are still in service
lest Saturday night and holidays
be their accusers. But what of work,
a drug beyond addiction?
'I continue to compose because
it fatigues me less than resting.'
Who could face meeting such a God
in abstract dark?
                    Grave lines of souls
are pressing to the waterside,
the boys like Botticelli angels
in the tinted air, the girls immaculate
in skinshine. How art nouveau
the very leaves of longitude
and berries of contentment!
A spider might swing from this
encampment of bold staves
in search of morning succulence.
                    The play-power oxymoron
has us in its thrall. We shall
never know how dense time was
in Paradise, nor what it is
to put things simply.

Meanwhile, the One
we nominate looks up from joy
of being here: he has not forgotten
what he wrote so many years ago
upon his pencil box:

John Everyman,
Home Villas, 1, Fabled Prospect,
Good Town, La Patria, This Hemisphere,
The Earth, The Solar System,
The Milky Way, The Universe.

Fast Forward

The view from Patmos, the ghost inside
the module! Living neither long enough
nor so curtly brings us nominative snakes,
time turned to blood upon the hour,
fours and sevens then the Lamb lies down
with Fury. And then Sir Headlong tells us
he has tripped and toiled along a donkey path,
the scala of so many angels, with a guide
who features in the brochures: old white hopes
that flavour an apocalypse, aunties
of the terrible abstractions, all made decent
when the fire has crept into a governed switch.

Now it is remaindered into visions,
into the history of the race, fast-forward,
the tape so stretched it might at any second
snap to oblivion. And they understood,
those heavy visionaries on poles
or sorting through the dung of lions,
that the countenance of state,
rock-featured and as bright as Caesar's eyes,
floats on insane shoulders. The boat for Patmos
is never late and though the island's rainfall's
doubtful its climate is compact despair,
angels in uniform identified on land.

Sitting with a little health-food lunch
among the flowers of a military estate,
we have to calm ourselves by crooning
death songs to the mush of midday heat.
Left to itself, the brain, circuiting the world,
becomes a rapid deployment force
and blasts ashore on any troubled sand.
Where will the end be staged, and whose hand,
held against the light, shall glow, brighter
than a thousand suns? Helicoptering
in lightning numbers, the god of prose
hears his own voice prophesying peace.

Cities of Light

Here, where as yet the minor monsters rove,
Yellow hats and orange jackets and
Theodolited stick-insects on tufts
Of spear-grass, our great ancestors put
One of the stations of the god of dreams,
Well-known to them as this bus-stop to you—
You cannot see it, though it's palpable.

And so with feral cats (a silly name
For constant friends) and tins and saucepans where
The sparrows and the mynahs splash, and dogs
That take what well-fed dogs refuse, we squat,
A hamlet of the harmless and the mad,
Defying parliament, that bank of acts,
Stung by the dew when Autumn tents us down.

The cops will come. This is a decent land,
So here injustice calls for little force
And fewer still will see it happening.
We can no longer hope for help from those
Who set the stars upon the pegs of night,
For they have gone away. But let none think
This metal world is newer than our own.

Ours was the great renunciation; we
Gave up everything that jumps and sparks
And flies. The curious who come to nose
At us will never leave the murderous weald
Of bulldozers, their kitchen-teatowel world
Of planted trees. They've sold their spirits to
Half-angels of some immortality.

The kind that speaks of progress and runs through
Every sacred thing like strings of fire,
In passage showing how a land may die.
Men do not live the longer and the rose
Will service them for graves as well as hearths.
Our people had these tools, our myths were mad:
We shrugged them off, we learned to live by light.

We took technology and turned it in
Upon itself—the grand computers learned
That trees had memory banks, that pollen was
A pure retrieval system and when ants
Assembled city states diplomacy
Attained a grace which even schools of night
Might envy, seeking Martian metaphors.

Our learning told us that the gods came down
Not to impose their rule on croft and creek
But to take holidays away from dark
In this domain more beautiful than theirs,
And like all tourists to seem out of place,
A nuisance to the locals, but with power
To make or mar their taut economy.

After repudiation came the dreams!
We drifted back to such complexity
Our daily actions seemed slow-motion rides.
When we were found by later schematists,
We looked in outward far from what we were,
Old Primitives, mere gatherers, un-wheeled,
Becalmed by symbols, unconcerned by time.

Our cities were invisible to them
And always will be while their eyes unhorse
Reality with expectation. This site,
Where even now the sub-contractors start
To plan their glass and galvanizing push,
Was never one of our more sterling forts,
A mere sub-station on the grid of dreams.

But worth defending. When they move us on
Their blindness will increase. One day perhaps
Their quondam cities filed away in squares
Will show our country's faces to the moon
Enacting open nothing to despair
And each surveying colonel will be hauled
Down from his statue in the seasoned park.

And then the true, the cities made of light,
Which once we had and which we gave away
Will come into their own. Already they
Are blazing in the night. Your astronauts
High-riding on Australia see their glow
And look at maps and read just emptiness
Outshining all the rules of blood and steel.